Dinosaurs Behaving Badly

Other books by Jason P. Schein and Jason Poole:

A Dynasty of Dinosaurs

Dinosaurs Behaving Badly

By Jason P. Schein

Illustrated by Jason Poole

Artemesia
Publishing

ISBN: 978-1-951122-33-1 (paperback)
LCCN: 2021952096

Cover Design: Geoff Habiger

Printed in the United States of America.

Artemesia Publishing, LLC
9 Mockingbird Hill Rd
Tijeras, New Mexico 87059
www.apbooks.net
info@artemesiapublishing.com

Dedication

We dedicate this book to our children; Logan, Izzy, Jackson and Lilly... May your lives forever be colored by science and nature.

And to the author's grandfather, Martin Schein, whose adventures in ethology helped to inspire a similar fascination and deep love of nature.

"Paleontology is the culmination of, and a perfect match for, a life full of curiosity and love for the natural world."
~ Jason Poole & Jason P. Schein

Introduction

I think every paleontologist would agree that we are constantly amazed by how much the fossil record can tell us. You might think that such an incomplete picture—one consisting almost exclusively of partial bones and teeth—would be so severely lacking in details that understanding these animals must be impossible. And yet, a more complete picture of their lives comes into focus every day. We can tell how fast dinosaurs grew and how long they lived. We can take their temperatures and measure how fast they walked and ran. We can even see what they ate, who was related to whom, and what made them go extinct.

But there is much more to dinosaurs and their lives than just their biology. What makes them both fascinating *and* fun—at least to us—is their behavior. How did they cooperate and communicate with their family members or with other members of their species? How did they compete with their enemies, obtain food, find mates, and interact with their environment? These questions can't be answered with bones alone. They also require imagination, creativity, and a lot of help from some more modern friends.

The best part about paleontology is that if we want to truly understand how these ancient animals may have lived so long ago, we must—no—*get to* study animals alive today. **Ethology**, or the study of modern animal behavior, helps us make sense of the dizzying variety of dinosaurs—now numbering well over 1,100 species with new ones named every month!—and the planet they inhabited. Paleontology is the best of both worlds: immersing ourselves in the beauty and complexity of nature to help us better, more completely understand the fossilized remnants of past worlds. *Really*... what could be better than that?!

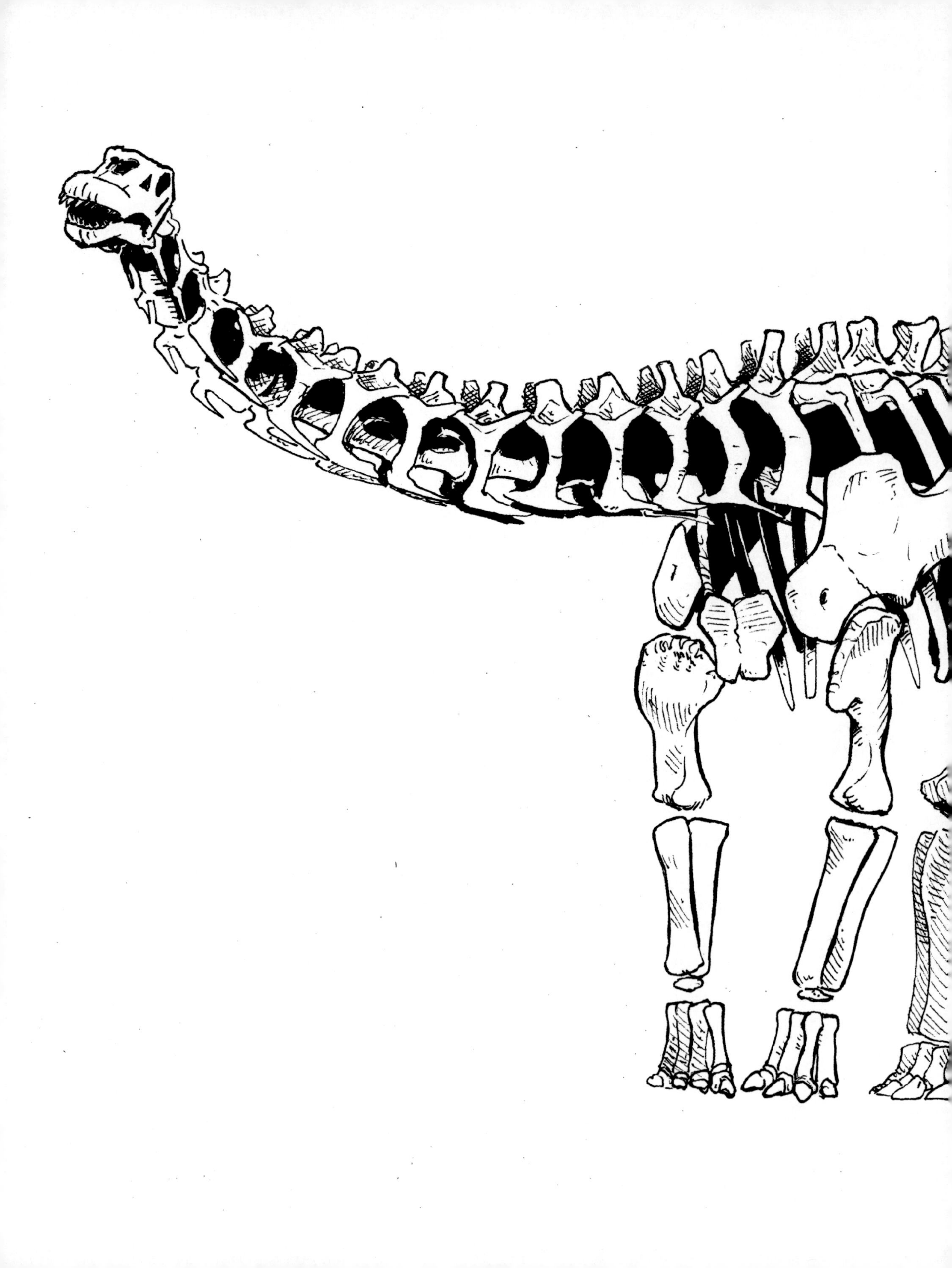

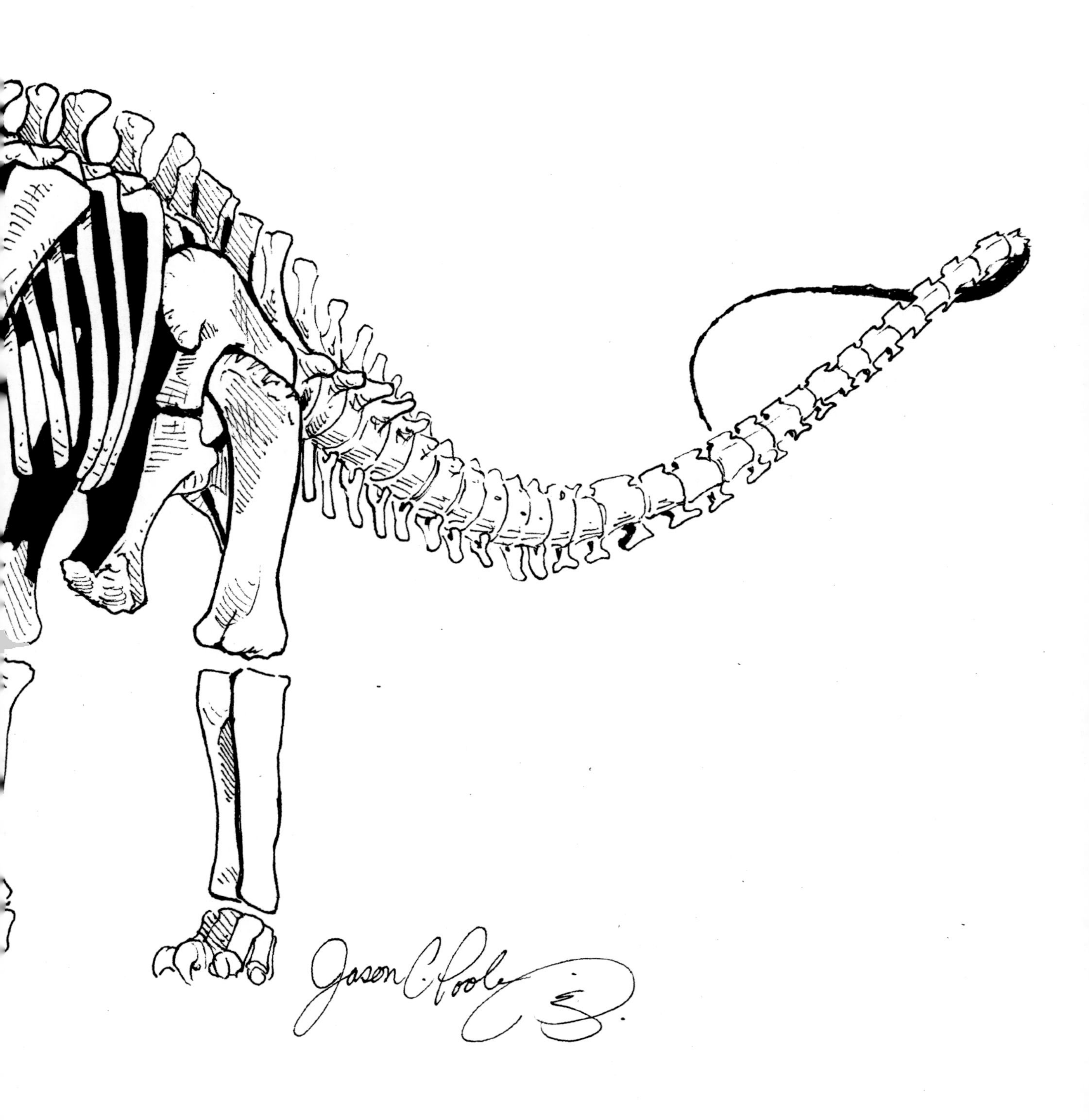

Fittest Rules

There are many rules in nature, but the first and most important is this: survival of the fittest. That doesn't mean the animals that hit the crossfit gym and slam protein shakes will be the ones that thrive. It means those individuals, and those species, that are best adapted to their environment will be the most successful. In nature, "success" means those animals get to breed, passing along their genes to a new generation, helping their species to persist.

It's important to remember, though, that species that are perfectly adapted to their environment today may not be in the future. That's because the environment, including the climate, geography, and the other plants and animals that share the **ecosystem**, are always changing. Change is inevitable, and to survive, a species must change with its surroundings, slowly, over long periods of time. In other words, they must evolve!

With three huge horns on its face and an enormous shield protecting its neck, *Triceratops* is one of the most recognizable and famous dinosaurs. It is also one of the most common dinosaurs, with hundreds of specimens filling museums around the world. Just like countless species living today with similar headgear, like deer and bighorn sheep, *Triceratops* probably often engaged in battles, not just with predators, but with other members of their species, herds, or families. So often these **intra-species** fights were over food, mates, territory, and dominance. In other words, they were fighting to establish who was the fittest.

Image:
Herrerasaurus
Late Triassic Period
(231 - 225 million years ago)
South America

Jason C. Poole 2021

Herrerasaurus

Image:
Triceratops
Late Cretaceous Period
(67 million years ago)
Western North America

Evidence of Feathers

The fossil record is astoundingly complete, but exactly *what* is preserved can be equally frustrating. When paleontologists talk about finding a new species of dinosaur, almost always what they're really talking about are the hard parts: the bones and teeth. Even those are hard to find, though, with complete skeletons being extraordinarily rare! The rest of the animal, including the skin, muscles, and organs, decay away and disappear in a matter of weeks or even hours. But if that is true, then how do we know that some dinosaurs had soft, delicate feathers?

In very rare cases, a dinosaur's skin, and anything attached to it, may have left an impression in the mud after it was buried, but before rotting away. Even more rare is when feathers themselves are preserved, often after sinking to the bottom of an ancient lake, or when trapped in amber. Paleontologists have even found evidence of feathers on dinosaur bones themselves, without actually seeing the feathers. Modern birds have structures on their arm bones called **quill knobs** which are the attachment sites for flight feathers. Paleontologists studying the fossils of *Velociraptor* found similar structures on their arm bones—a great example of using modern species to help us learn more about ancient ones! By making a careful examination we know that *Velociraptor* had feathers, even if we haven't found feather impressions in the fossil record.

But what exactly are those feathers for, anyway?

Image:
Deinonychus
Early Cretaceous Period
(115-108 million years ago)
Western North America

Velo-

Image:
Velociraptor
Late Cretaceous Period
(75-71 million years ago)
Asia

Preening is Cleaning

Feathers come with a lot of advantages. They're great **insulators**, of course, keeping their owners both warm and dry. They also provide a platform with which they can communicate to predators, rivals, or potential mates, through an endless variety of both colors and patterns. They also happen to be excellent, lightweight tools that help birds rule the skies! But all those advantages also come with some drawbacks. All of those feathers can get dirty, too dry or oily, and certainly provide endless hiding places for parasites. Any of these factors can affect the usefulness and efficiency of the feathers, and even the health of the bird.

Birds spend a great deal of time **preening**, or cleaning and maintaining their feathers with their beaks. Based on the behavior we observe every day in modern birds, we know that feathered dinosaurs, like *Gallimimus*, must have preened themselves as well. Preening repositions feathers and realigns their filaments, which improves their aerodynamic and insulating properties. It also keeps feathers clean and dry, and even removes those pesky parasites. Some birds have even incorporated preening into their social behavior. Did dinosaurs preen their mates? Did they have a gland to secrete oils onto their feathers, like birds living today? So far, we can only guess, but it's certainly possible!

Image:
Gallimimus
Late Cretaceous Period
(70 million years ago)
Asia (Gobi Desert, Mongolia)

Visual Communication

In the animal kingdom, it is very rare to have a body part or behavior that is used for just one thing. A frog's long legs are great for swimming and leaping great distances, as we all know, but they also are a convenient length for cleaning their bodies and heads. Feathers are great for flying, keeping warm and dry, and for creating colorful displays, and antlers are a great way to show off and make great weapons for battling rivals. Nature may have endless imagination, but it's also quite efficient, finding multiple uses for almost every structure and behavior.

Everyone knows that the thagomizer—the spikes on the end of the tails of stegosaurs—was used for defense against predators like *Allosaurus*. But what about the plates on their backs? Early paleontologists thought they were for armor, but the more we've learned about stegosaur skeletons, the less sure we are that this was the case. The plates are just modified **osteoderms**, or bones that grow directly out of the skin, and they're also rather fragile—neither condition is very useful as armor. Maybe instead they were used for communication? Stegosaurs may have used bright colors on the plates to tell other stegosaurs they are ready to mate, that danger is nearby, or just to signal it's "mood" like chameleons do today. Or maybe they used these plates to make themselves look bigger to potential attackers than they really were—a form of defense, but through communication rather than combat. So, which is it—defense or communication? Knowing how efficient Mother Nature and the animal kingdom is, probably both!

Image (previous):
Stegosaurus
Late Jurassic Period
(152 million years ago)
Western North America

Rutting

During mating season, animals of all types do some pretty weird things—*especially* the males. Male **ruminants**—the group of herbivorous, hoofed mammals that include deer, antelope, and goats—mark their territories with visual cues made by scraping their antlers or horns on trees. Some also mark their territories with scent, either by rubbing special scent glands on trees and dirt, or by urinating on them! Bull elk herd their females away from other males, and bull moose are known to fight bitterly to establish dominance—sometimes to the death! Rhinos, Hercules beetles, hornbill birds, and walruses are just a few species alive today that subscribe to the "bigger is better" theory when it comes to showing off to attract mates.

We don't know how *Pachyrhinosaurus* behaved during their mating season, or any other dinosaurs for that matter. This is one type of behavior that is unlikely to leave any kind of trace in the fossil record. But, finding and courting mates is nearly universal in the animal kingdom, so we have to study nature and all the different ways animals conduct this important behavior to interpret how dinosaurs may have behaved. *Pachyrhinosaurus'* horns and frill were certainly useful as a defense against predators, but just about any modern animal with similar structures also uses them for competition for mates. Could they have used them for display? Could they have flared them into more flashy colors or positions, like many birds do? So far, we can only guess.

Image:
Pachyrhinosaurus
Late Cretaceous Period
(70 million years ago)
Western North America

Jason C. Poole 2021
Centrosaurus/Pachyrhinosaurus

Herding

Lots of different animals live together in large groups. Birds gather in flocks, dolphins dive in pods, and fish swim in schools. The reasons for this behavior are pretty simple: safety in numbers. If you're a relatively defenseless animal, usually an herbivore, like a zebra or a gazelle, being in a large group makes it easier for you to avoid predators. The group *is* your defense. The opposite is true, too. Humpback and killer whales and many other predators are more successful when they hunt cooperatively.

One drawback to living in large groups is that the food or water sources in an area are quickly exhausted. The herd must be constantly on the move to find enough to eat. Many of Africa's modern **megafauna** still migrate, and not that long ago, hundreds of millions of bison were perpetually roaming en masse across great swaths of North America.

We don't know exactly how far back in the history of life on Earth herding got its start, but it is very clear that many dinosaur species lived in herds. *Styracosaurus*, *Diplodocus*, and *Protoceratops* have all been found in bone beds containing dozens—or hundreds—of individuals that died together. *Maiasaura* and *Oviraptor* both nested in large groups, just like many birds do now. And now, based on **gastroliths**, or stones picked up and swallowed to help digest their food (just like birds do today!), we even have evidence that some dinosaurs, like *Apatosaurus*, migrated from present-day Wyoming to Wisconsin and back, 150 million years ago!

Image:
Apatosaurus
Late Jurassic Period
(152 million years ago)
Western North America

Flocking

There are plenty of animals that live almost their entire lives as solitary, isolated individuals, seeking out others only for short periods of the year in order to find a mate. But as we have already seen, many other species are very social, living in large family groups or even crowds with seemingly endless numbers of individuals. Many species of fish live in large schools, and for sardines in particular, those schools can contain millions of individuals. The largest single group of mammals ever witnessed couldn't be counted, but it took three entire days for that herd of springbok to travel through southern Africa in the mid-1800s. Birds, too, can gather in extraordinary numbers. North America's passenger pigeons are said to have numbered five *billion* individuals and migrated in flocks that would block out the sun for hours at a time, before they were hunted to **extinction** in the early 1900s.

Flocking is the behavior of birds as they eat or travel in a group, and since theropod dinosaurs are birds' ancestors, it is probably the best term to use for a group of these meat-eating dinosaurs. Paleontologists don't yet have direct evidence of flocking among theropods, but since the behavior is so common among living birds, they aren't going too far out on a limb to suggest that it probably did occur.

Image:
Struthiomimus
Late Cretaceous Period
(70 million years ago)
Western North America

Jason C. Poole

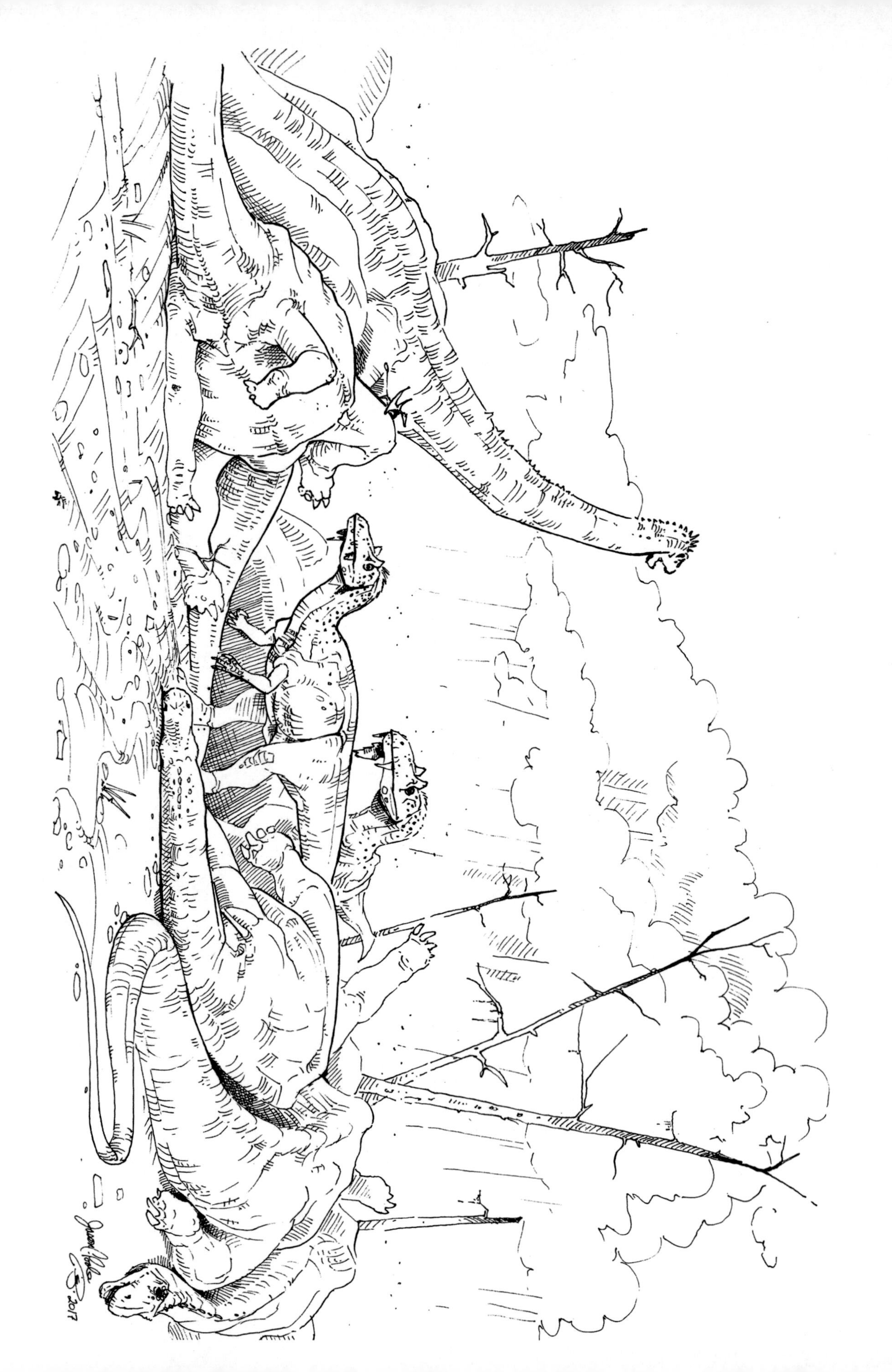

Herding by Age

The "Mother's Day Quarry" is one of our favorite stories because it is a site we have worked in for several years, and because the site has so much to tell us. This dinosaur **bonebed** has been worked by three different crews, including ours, over 15 seasons. In that time, almost three *thousand* bones have been found in an area that's probably smaller than your front yard. That's an *astounding* abundance and density of fossil bone, but what is even more interesting is that almost every single one of those bones is from a *Diplodocus*. But the story gets even more interesting, because all of those bones are from baby or juvenile *Diplodocus*—not a single adult!

The paleontologists who worked this site before we inherited it developed their own insightful theories to explain this unusual occurrence. Their best guess is that a herd of young *Diplodocus* were traveling together and huddled around a usually reliable watering hole during the dry season. As this drought persisted, though, they died together, succumbing to the heat and dehydration. If true, this scenario suggests that *Diplodocus*, and perhaps many other dinosaur species, traveled in age-segregated herds—a fascinating insight into the behavior of dinosaurs that we could never have deduced based on their bones alone!

Image (previous):
Diplodocus
Late Jurassic Period (152 million years ago)
Western North America

Image (following pages):
Diplodocus
Late Jurassic Period (152 million years ago)
Western North America

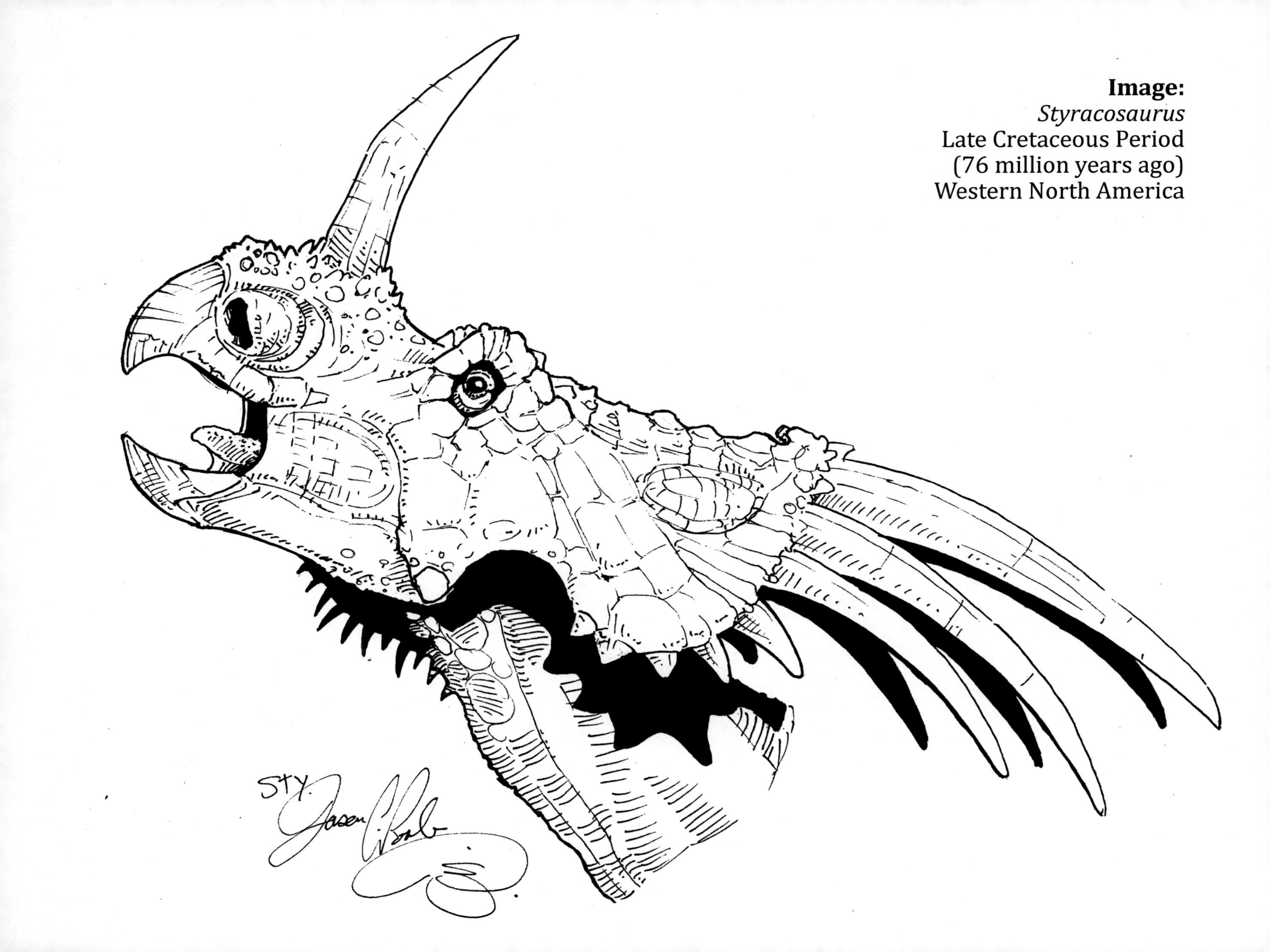

Image:
Styracosaurus
Late Cretaceous Period
(76 million years ago)
Western North America

Horn Variation

The Ceratopsians, or horned-face dinosaurs, seem to be a particularly diverse group, especially in what is now western North America. By some estimates, more than 30 new species were described from 2010 to mid-2015 alone—that's more than 1 every two months! They all generally have the same, four-legged, bulky body plan, so one of the best features we can use to easily tell them apart, other than size, is the shape and arrangement of horns and shield on their heads. The sheer variety is sometimes overwhelming and difficult to understand, until you think about all of the antlered and horned animals living today.

There are hundreds of species of animals living today that announce their presence with elaborate headgear, and the variety of shapes and styles among those bony projections is equally astonishing. From the tiny pronged horns of the diminutive dik-dik and the palm-shaped antlers of a massive moose, to the coiled horns of bighorn sheep and the spiraled spears of the oryx, these incredible structures remind us that nature's imagination is boundless!

Mystery Arms Solved

The history of the discovery of *Deinocheirus* is one of the strangest in all of paleontology. In the 1960's scientists found the shoulders and arms of a new dinosaur in Mongolia's Gobi Desert. They named the dinosaur *Deinocheirus*, meaning 'horrible hand,' believing them to be the powerful claws and arms of a ferocious predatory dinosaur. For almost fifty years, nothing else was ever found or known about this dinosaur—just that single, nearly perfect set of forelimbs.

Finally, two new specimens of *Deinocheirus* were found, but each had portions stolen by poachers only a short time before. A single bone missed by poachers at one of those sites was later used to track down the looted remains, but only after they made it around the world on the black market—an amazing bit of detective work! Once the bones were reunited in 2014, paleontologists finally had two nearly complete skeletons, and a completely different understanding of this amazing dinosaur. We now know it was a bizarre, omnivorous ornithomimosaur, or "ostrich-like" dinosaur.

The question remains, though: what were those long arms for? Paleontologists' best guess so far is that maybe those long arms and claws were for digging. Anteaters may be missing the long arms, but they have similar claws used for demolishing termite nests. *Deinocheirus* is also thought to have eaten fish—could those long arms have been used to grab slippery aquatic critters too?

Image:
Deinocheirus
Late Cretaceous Period
(70 million years ago)
Asia

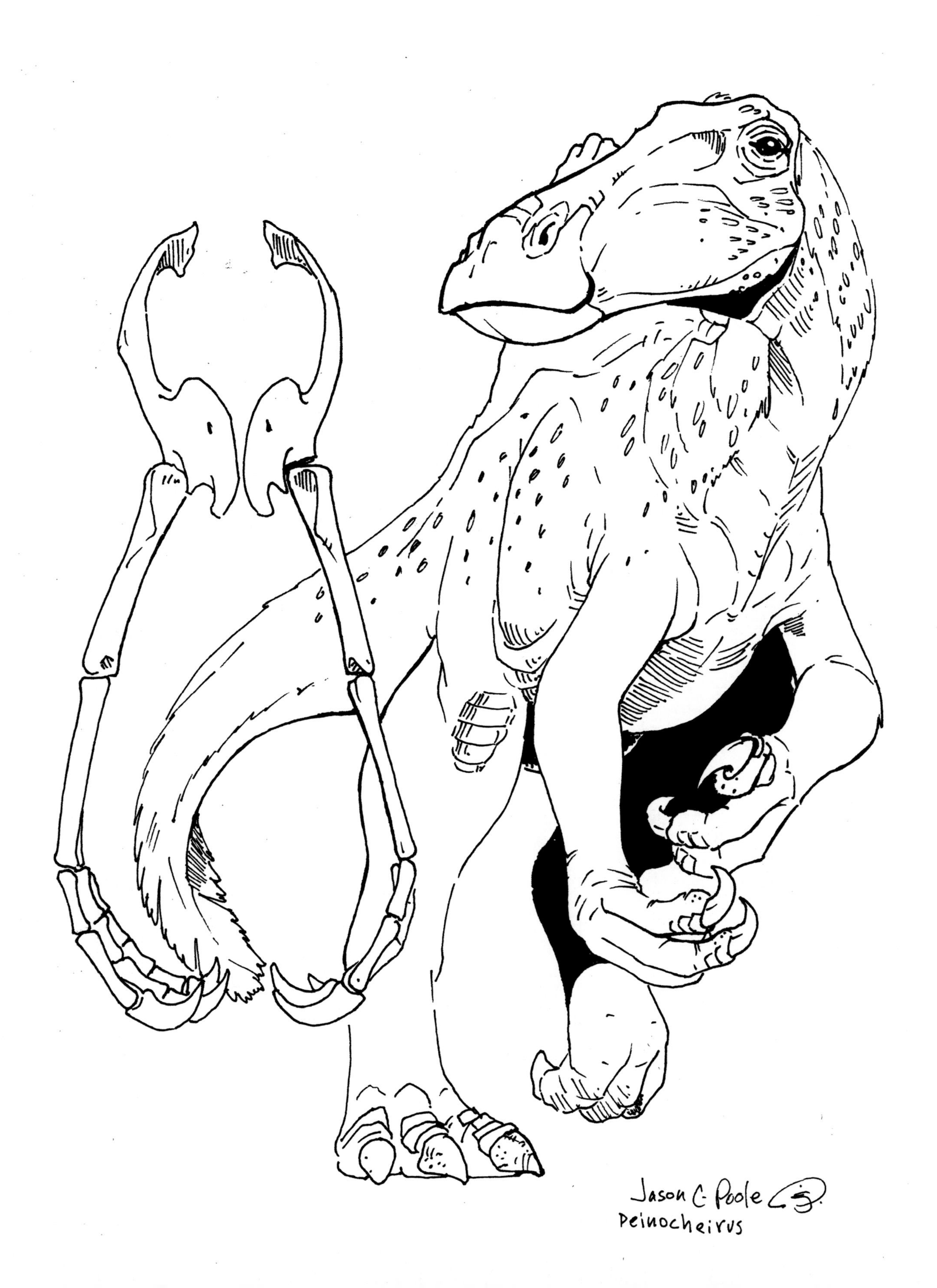
Jason C. Poole
Deinocheirus

Beak function

Beaks are fascinating structures. They come in all shapes and sizes—think toucans and flamingos, spoonbills and pelicans. But it's not just birds that sport these useful structures: turtles, some tadpoles, pufferfishes, and squids all have beaks. The one thing all these various beaks have in common is that they are made of **keratin**—the same material that makes your fingernails and hair.

Many dinosaurs had beaks too. This isn't surprising since we now know modern birds evolved from the meat-eating dinosaurs like *Limusaurus*. What may surprise you, though, is that plenty of other dinosaurs had beaks, too. The ceratopsians *Leptoceratops* and *Triceratops* sported their own mouthguards, as did some of the two-legged herbivores, like *Hadrosaurus* and *Iguanodon*. But how can we tell what the beaks were used for?

All those different beak shapes aren't just for looks. Both dinosaurs and birds evolved distinctly different beak shapes for distinctly different jobs. In birds, some beak shapes are better for nimbly picking fruit from trees, whereas others are better for crushing hard seeds and nuts. Some bills are more adept at snatching quick, elusive prey, or sucking up nectar from deep inside a flower, and still others are perfect for sifting through sediment under water for tiny invertebrates. By observing beak shape and function in modern birds, paleontologists can make informed guesses about what dinosaurs with similarly-shaped beaks ate.

Image:
Iguanodon
Early Cretaceous Period
(124 million years ago)
Europe

Jason C.Poole

Burrowing

Paleontologists discover new dinosaur species all the time—on average about one every ten days! But it's not all about new species—once in a while we find a dinosaur occupying a new **niche**, or a specific role within its ecosystem. It's only in recent years, for example, that we found dinosaurs that excavated and lived in burrows. We now have evidence that several dinosaurs, including *Orodromeus* and its cousin, *Oryctodromeus*, excavated and lived in their own underground lairs.

Burrowing is one of the most common and cozy ways modern animals make their homes. Burrows not only provide safety from predators, but are also dry, cool in the summer, and warm in the winter. We've all seen chipmunks peaking their heads out of holes or between rocks, but some burrowing animals might surprise you. There are several species of burrowing birds, like puffins in the Arctic and owls in North America's prairies. Foxes seek protection in their underground dens, and gopher tortoises slowly dig burrows in the southeastern U.S. that can be almost 50 feet long! Even animals as big as bears live in underground dens, especially for hibernation in winter. Prairie dogs truly are **subterranean** engineers, burrowing cooperatively to construct elaborate cities underground! Their burrows may have 6 entrances, have different rooms for specific purposes, and even have their own natural ventilation systems.

Image:
Orodromeus
Late Cretaceous Period
(77 million years ago)
Western North America

Jason C. Poole
Orodromeus

Image:
Parasaurolophus
Late Cretaceous Period (76 million years ago)
Western North America

Vocalizing

Making sounds is probably as common a method for communication throughout the animal kingdom as patterns of color and is just as varied. We've all heard dogs bark and birds sing, of course, but it goes much further than that! Whales sing songs that are answered by other whales thousands of miles away. Dolphins use sound to maintain their pod relationships, but also as sonar to locate and hunt food. Howler monkeys famously howl at deafening volume, in part to warn potential rivals that this is 'their' territory. Of course, the most universal reason to vocalize is to find a mate. Many animals that are generally silent most of their lives, like male elk, are well known to be very vocal during the fall **rut** while in search of females. And have you ever heard the mating calls of Birds of Paradise in Papua New Guinea? It's like something from another world!

Paleontologists may never find evidence of vocal cords in dinosaurs—they are, after all, those soft parts that decay so quickly after death—but that's no reason to assume dinosaurs weren't vocalizing just as often, and for all the same reasons, as animals living today. At least one dinosaur did generously leave us with evidence that it was actually quite noisy: *Parasaurolophus* sported a very distinctive crest on its head, which we've since discovered was actually a series of hollow tubes, just like the tubes in many musical instruments. It is very likely that, at times, the Era of the Dinosaurs was quite a noisy place!

Pair Bonding

Nature, through trial and error and over billions of years, has developed many different strategies to help species reproduce and survive over geologic time. A strategy that works well for humans and killer whales may not be what's best for mountain lions and toads.

Pair bonding is a behavior in which two individuals of a species mate for life, and it is not as uncommon across the animal kingdom as you might think. Voles, wolves, and beavers have life partners, as do some primates, like gibbons, and even some species of fish. Birds, especially, tend to pair bond, with bald eagles and penguins being among the most famous, committed partners.

There are also, of course, plenty of examples of the opposite strategy as well. Adult males and females of many species, like mountain lions, meet only long enough for courtship and mating, and then have no further interaction, leaving all of the parental duties to the mother.

We don't know when pair bonding first evolved, and there's no way—*yet!*—to be certain whether or not some species of dinosaurs sought life mates, but it seems entirely likely. Could *T. rex* fathers and mothers have been nurturing, caring partners? We may never know, but it's possible!

Image:
Ceratosaurus
Late Jurasic Period
(161-145 million years ago)
Western North America

Jason C. Poole

Ceratosaurus

Image:
Tyrannosaurus rex
Late Cretaceous Period
(66 million years ago)
Western North America

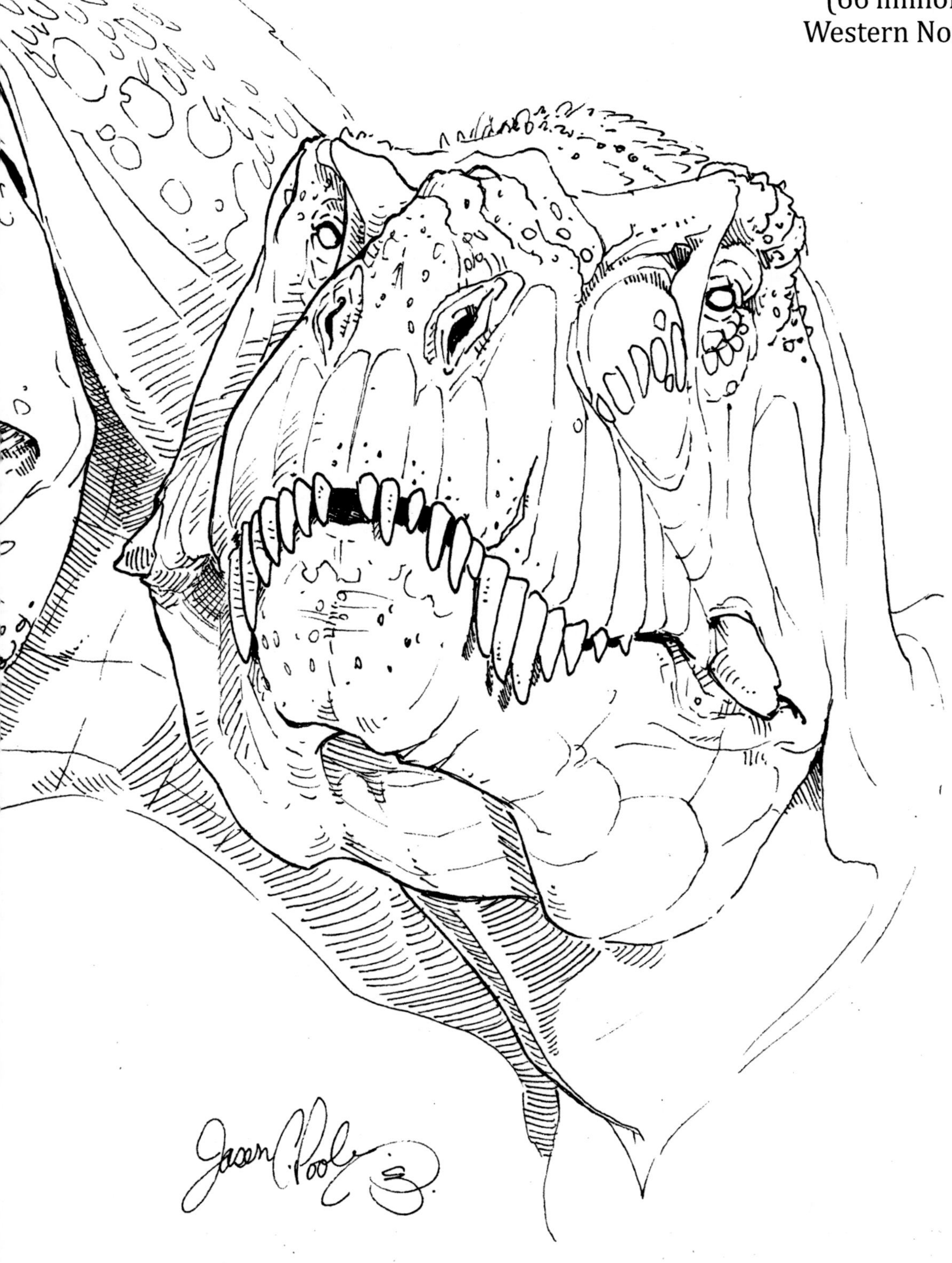

Bath Time

All animals must take the time and effort to keep themselves clean. It's not just about their looks—cleanliness is a matter of life and death! To stay healthy and avoid infections, many animals must keep their skin, fur, or feathers free of dirt, mud, and parasites. A great way to do that is to go for a swim!

As we have seen, birds must keep their feathers clean to maintain their aerodynamic and insulating properties, so an important part of their preening routine for some species is to take a dip in a puddle. Other animals aren't too picky about the water they bathe in. Elephants, rhinos, and pigs love to wallow in muddy water. Water that looks filthy to us still helps to rid them of parasites but has the added benefit of acting like sunscreen. Of course, there are also some animals that hate water but still manage to keep themselves clean. Many of the big cats, like lions and cheetahs, *hate* being wet, but still bathe by taking "cat baths," or licking themselves all over, to keep their coats shiny, clean, and parasite-free. No matter which method works best, you can bet that dinosaurs used some of these same strategies.

Image:
Bistahieversor
Late Cretaceous Period
(75 million years ago)
Western North America

Jason C. Poole
Bistahieversor

Cooling Off

As you well know, getting and staying clean is not the only reason for taking a bath or going for a swim. It's also a great way to cool off! You think summer is hot now? For much of the Mesozoic Era—the Age of the Dinosaurs—taking a cooling dip must have been an important, and sometimes lifesaving, strategy for keeping their cool.

Dinosaurs alive during the Late Cretaceous Period, in particular, lived through the hottest climate Earth has ever experienced. Levels of greenhouse gases in the atmosphere, the same heat-trapping gases causing climate change today, were at all-time highs, creating temperatures so high that there was no ice even at the north and south poles! This helped to create sea levels that were more than 300 feet higher than today. It's difficult to imagine, but under those warm and humid conditions, dinosaurs thrived even in Antarctica—a place where reptiles could never survive today!

Plenty of animals living today will take a nice cooling dip when the shade just doesn't cut it. Cattle do this often, and hippos practically live in the water, in part to avoid the harsh African sun, coming on land at night to feed. Even elephants and pigs take baths too cool off, though usually those are mud baths!

Image:
Hadrosaurus foulkii
Late Cretaceous Period
(80 million years ago)
Western North America

Dust Bath

There's one more way some animals like to take a bath, and it may be one you've never tried before. Dust baths are exactly what they sound like—rolling around in dry dirt to rid themselves of parasites on their skin, fur, or feathers. Many animals, from chickens to horses, and elephants to prairie dogs, enjoy a good roll in the dirt to help rid themselves of ticks and other pesky critters. Some animals also use this as a way to mark their territory with their scent. Like chickens and other dust-bathing birds, *Bambiraptor* may have fluffed up its feathers and spread its wings to try to get as much dirt all over its body as possible.

American bison famously love a good, dusty dirt bath. Scientists have noticed that many individuals will use the same spot, rubbing it clear of vegetation and creating a small depression in the ground, which tends to hold rainwater and snowmelt just a little longer than the surrounding prairies. In an otherwise dry landscape, those extra days of moisture make all the difference, allowing for the survival of a whole suite of different plants and creating important micro-ecosystems. So in this way, bison are ecosystem engineers. If 3,000 lb bison can do all that, imagine what a 130,000 lb *Giraffatitan* dust bath could do!

Image:
Bambiraptor
Late Cretaceous Period
(72 million years ago)
Western North America

Jason C. Poole
Bambiraptor

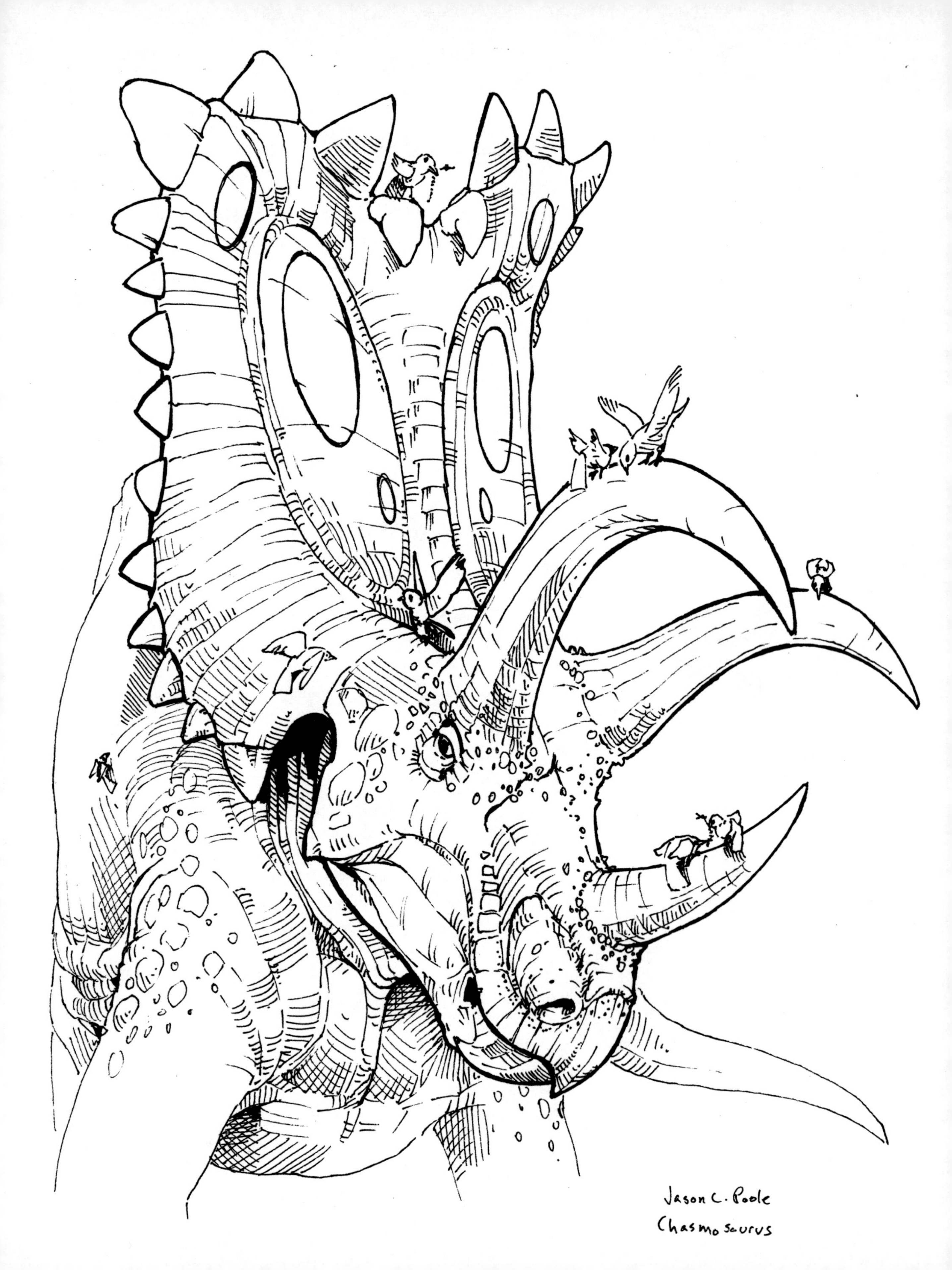
Jason C. Poole
Chasmosaurus

Pesky Parasites

Some animals don't seek out water or dirt to rid themselves of parasites. Instead, the bath comes to them! Oxpeckers, for example, are African birds that live their whole lives on the backs of giraffes, rhinos, and other large mammals, grazing on ticks, insects, and even larvae plaguing infected wounds. This is a classic example of cleaning **symbiosis**: a relationship between two species in which one species eats the parasites off of the other. There are several species of fish and shrimp that clean other, larger fish, and warthogs are sometimes cleaned by banded mongooses. Loggerhead sea turtles even get cleaned by a species of crab.

There's no way to know if any species of dinosaurs enjoyed a similar arrangement with live-on maids, but it seems like a good idea. Parasites have plagued other animals since, well, there have been animals. With such enormous bodies, big head shields, or tiny arms, there certainly were plenty of hard-to-reach places for ticks, fleas, and other vermin to hide and make nuisances of themselves. Without the help of a friendly cleaning service, it seems unlikely that those enormous bodies could have stayed clean and healthy.

Image (previous):
Chasmosaurus
Late Cretaceous Period
(75 million years ago)
Western North America

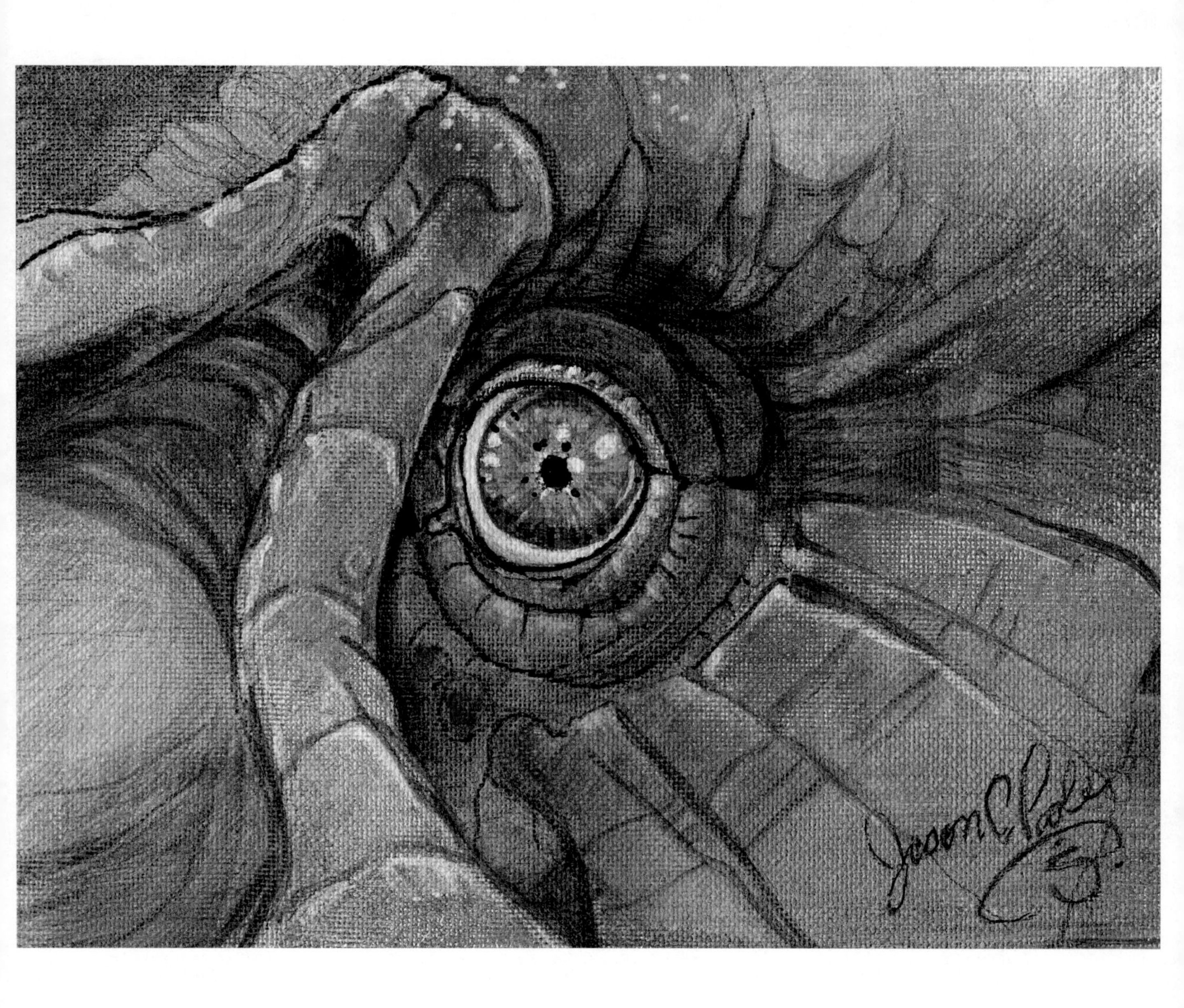

Image:
Tyrannosaurus rex
Late Cretaceous Period (66 million years ago)
Western North America

The Better to See You With

Astounding discoveries seem to happen quite often in paleontology. It seems that every year researchers find something that we previously believed to be entirely impossible. For example, it wasn't long ago that we thought finding patterns of pigments in dinosaur skin and feathers was completely impossible, but recent discoveries suggest they may be real! Sadly, though, there is a limit to what we can reasonably hope to learn because the fossil record is so incomplete, and fossilized soft tissues are exceptionally rare. Mineralized bones and teeth are tailor made for fossilization, but soft tissues like muscles, organs, and eyes, decay far too quickly to be preserved in the same way. It's difficult to imagine that we'll ever be able to study the internal anatomy of a dinosaur's eyeballs.

That's not to say we can't make educated guesses about their sight. By studying modern animals, we know that different species see in different ways. Many snakes see in **infrared**—which means they see patterns of heat naturally given off by their prey, rather than the colors of those animals as we do. Many insects have compound eyes made up of thousands of tiny lenses, and dragonfly brains work so quickly that they actually "see" in slow motion! Birds are often believed to have eyesight that is far superior to our own. Structures in the eyes of birds allow them to see a much more vibrant world, filled with more colors that we can only imagine. Can you guess which bird is thought to have the *BEST* vision of all? The lowly pigeon!

Nocturnal Hunting

We all know animals that are well-adapted for nighttime activity. Bats use **echolocation**, or sound waves to detect their prey. Owls utilize incredible hearing, eyesight, and silent feathers to find and surprise their prey. Some primates, like tarsiers, have truly enormous eyes to help them navigate their jungle homes at night; each of their eyes are as big as their brain! Being active at night isn't easy, but plenty of animals have found ways to make it work for them.

It is difficult to imagine that there weren't some dinosaurs equally well-adapted for an active nightlife. It's safe to assume that just like those modern animals, nocturnal dinosaurs had excellent eyesight, likely aided by extremely large eyes. Some also may have had color patterns that would allow them to blend into the forests' dark shadows, either to avoid being eaten, or to more easily sneak up on potential prey.

It's unlikely we'll ever discover fossilized eyeballs, but certainly the size of the opening for the eyes in the skull can give us clues to how good their night vision may have been. We can also analyze the size and shape of the **braincase**: if the size of the portion of the brain that controls and processes eyesight is larger than normal, it may very well have had excellent vision for nighttime hunts!

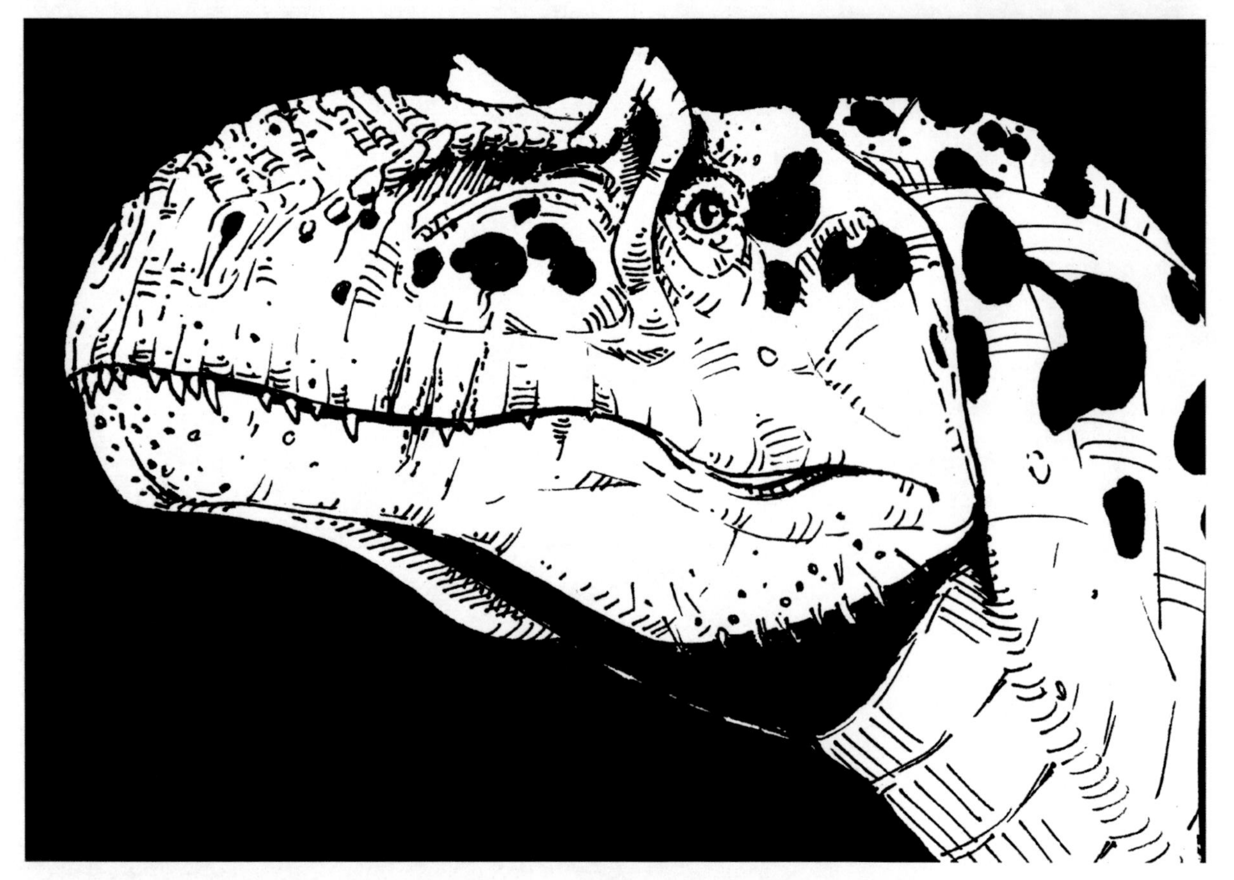

Image:
Cryolophosaurus
Early Jurassic Period
(190 million years ago)
Antarctica

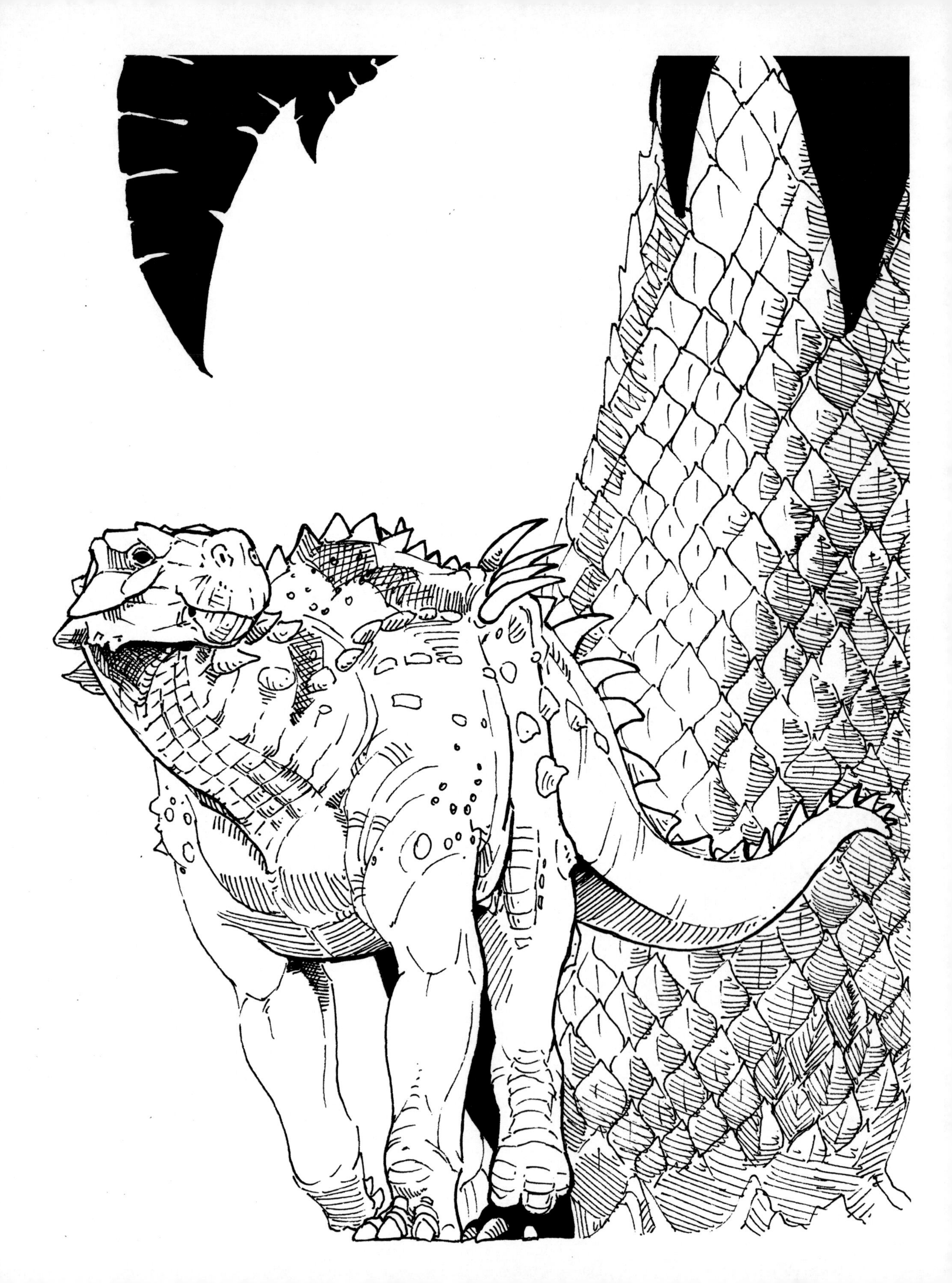

Armor

There are really only two types of defensive armor that animals use to protect themselves from predators: long, sharp spikes, or dense, hard plates that grow within the skin, or instead of the skin. Porcupines, of course, prefer the spikes, but plenty of other animals have adopted the armor plates. Armadillos are a classic example of modern-day walking tanks, but pangolins have a similar covering, and turtles took the concept of body armor to a whole other level. Some fish grow extremely thick scales, and crocodiles grow bony plates within the skin on the backs. Even rhinoceroses have some armor—their skin is both extremely thick and segmented, which allows for flexibility. In fact, this may be the closest thing in the animal kingdom to a Medieval knights' suit of armor.

Many dinosaurs evolved this defensive strategy as well. The most famous of these were the ankylosaurs, some of which even had both armor plating and large, formidable spikes. With all that extra weight, ankylosaurs never tried to outrun predators, but instead, relied on the armor to deter them. But they also had one more trick up their sleeves, and one we don't really see among any modern animals—huge bony clubs that they would swing back and forth. One solid whack with these giant hammers could easily break the legs of even the biggest predators, so surely most kept their distance!

Image (previous):
Minmi
Early Cretaceous Period
(125 million years ago)
Australia

Tails for Defense

Many sauropod dinosaurs were pretty much out of luck when it came to actively defending themselves against predators. Larger adult sauropods could rely on their overwhelming mass to discourage attack, much like adult elephants do today, but only when fully grown. Some species, like *Saltosaurus*, had osteoderms, or armored boney plates in their skin, but those probably weren't very effective against attack. There is another, more unusual strategy, though, that a whole group of sauropods used.

Suuwassea is a member of the **Flagellicaudata**, or 'whip-tailed' dinosaurs. These sauropods had extremely long, thin, and flexible tails that some paleontologists believe could have been used like whips. By some estimates, the ends of those tails could have reached 1,300 miles per hour, creating a snap as loud as the largest cannons on battleships!

As bizarre as that sounds, there are some animals living today that have a similar, if less loud, defense. Iguanas are known to use their tails as whips when threatened, as anyone who has ever owned one as a pet will tell you! And even that is not as bizarre as other tail-based defensive strategies. Some modern lizards are known to drop their tails when attacked by a predator. The lizard runs away to live another day and grow another tail, while the disembodied tail continues to wriggle! It's a bizarre sight, but can you imagine a 5 ton, 30-foot-long tail doing the same thing? There is no evidence any dinosaurs used that same strategy, but it sure is fun to think about!

Image:
Suuwassea
Late Jurassic Period
(150 million years ago)
Western North America

Jason C. Poole

Image:
Suuwassea
Late Jurassic Period
(150 million years ago)
Western North America

Feeding

All animals eat, of course, but there are many different ways animals consume their food. The tiny animals that build coral reefs, and clams that live in lakes both eat by filtering tiny food particles out of the water as they flow by. The largest animal that ever lived—the blue whale—also filters small animals as it slowly swims across the ocean's vast distances. Some sea urchins eat sand as they burrow through the ocean bottoms. Their stomachs digest any organic matter that happened to be there, and then they poop clean sand out the other end. The world would be a pretty dirty, smelly place, if we didn't also have scavengers, or animals that eat other dead animals. Wolves, sharks, and vultures are all scavengers, at least part of the time, and are vitally important to healthy ecosystems. And of course, there are those that actively hunt down their own prey for food. Like scavengers, predators, like polar bears, hawks, and chameleons, also help maintain healthy ecosystems by controlling the populations of other species. Predators even help to maintain the health of those prey species by preying mostly on sick or injured individuals.

Nature is a very complex web of interconnected species—every species, and whatever feeding strategy they use, affects all the others around them.

Image:
Acrocanthosaurus
Early Cretaceous Period
(115 million years ago)
Western North America

Jason C. Poole
Acrocanthosaurus

Browsing vs Grazing

Both browsing and grazing animals are **herbivores**, but it's exactly what kind of plants the animal eats that makes the distinction.

Herbivores that mostly eat the leaves, young branches, and fruits of taller, woody shrubs and trees, are **browsers**. Elk and moose are classic browsers in North America, but in Africa the most prominent browsers are elephants and giraffes. Some of the long-necked sauropods, like *Spinophorosaurus* and *Brachiosaurus*, are classic examples of upright sauropods that could only have enjoyed their dinners with a view, munching only the finest leaves at the tops of trees.

Grazers are animals that eat low growing plants, most often grass. Horses and cattle are grazers, but so are rabbits, hippos, grasshoppers, and even some sea turtles—green sea turtles graze on beds of seagrass!

Diplodocus, *Nigersaurus*, and other grazers in the era of the dinosaurs had plenty to eat, but it wasn't grass. Ferns, horsetails, and other shrubs were their favorite foods because they were abundant and could be found almost everywhere, and because grass didn't yet exist. Grasses first evolved in the latest Cretaceous Period and were encountered—maybe even eaten—by only the very last dinosaurs, like *Triceratops*.

Image:
Diplodocus
Late Jurassic Period
(153 million years ago)
Western North America

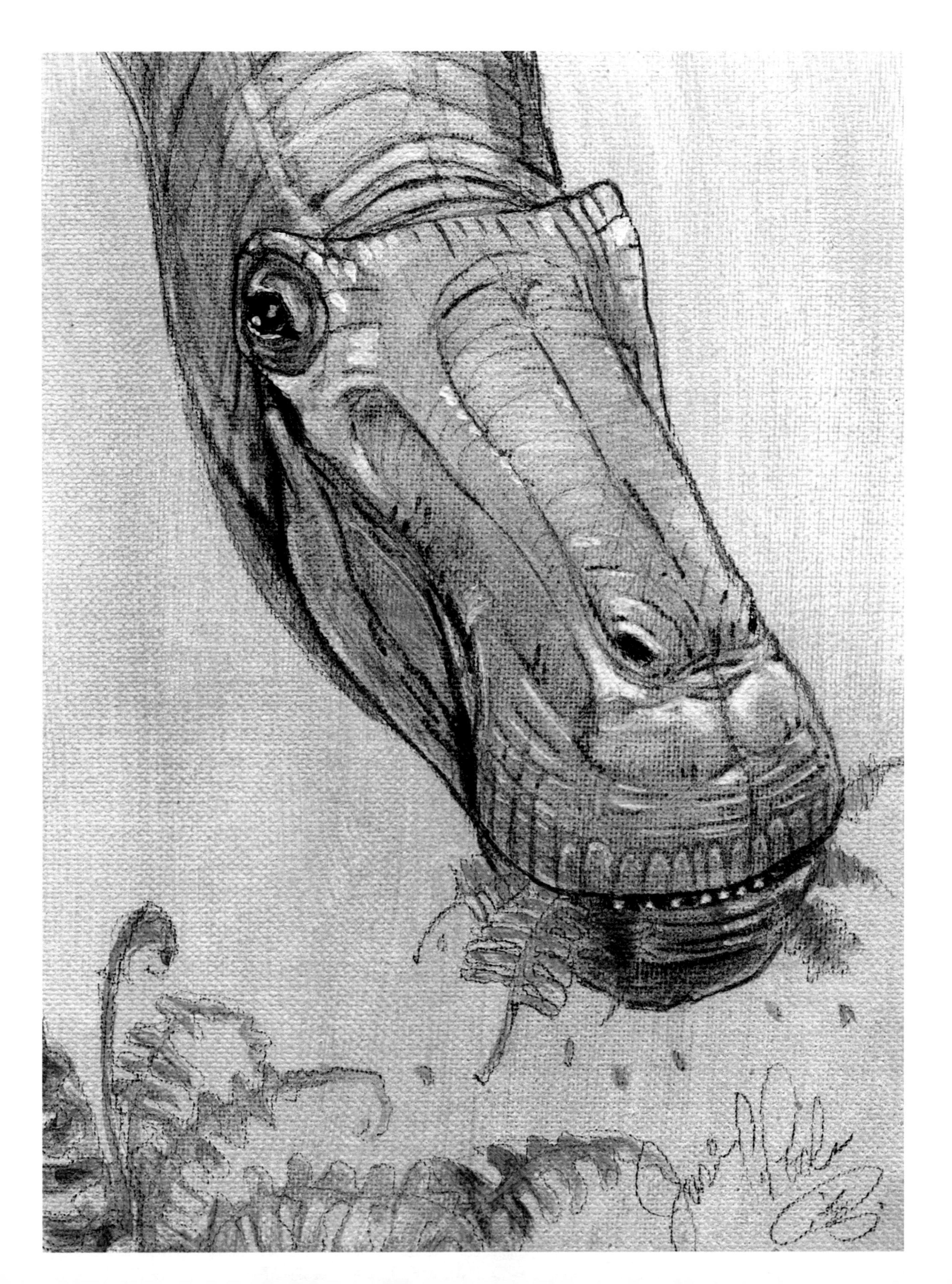

2021
Tyrannosaurus rex

Hunting vs. Scavenging

For animals that prefer some protein at every meal, there's really only two ways to get it: kill it yourself or let something else kill it first.

Scavengers prefer to let someone else do the hard work, but they have a very important job. In fact, without scavengers to clean up **carrion**, or decaying animals, life as we know it couldn't exist! Nature *needs* dead plants and animals to be broken down and recycled back into the ecosystem. **Obligate scavengers**, or animals that exclusively feed on carrion, include many insects and vultures. Hunters, of course, take a more "hands on" approach to obtaining their meals by killing it themselves.

In nature today, there aren't very many animals that are exclusively hunters or scavengers. Well-known hunters like lions, wolves, and raccoons won't pass up a free meal that they come across, and plenty of animals that more often eat carrion, like hyenas, are also perfectly capable of hunting when they're hungry.

Animals living today tell us the same is probably true of the carnivorous dinosaurs. Some paleontologists suggest that *T. rex* may have been a scavenger, while others argue it was definitely a skilled hunter. In truth, *T. rex*, like many **carnivores** living today, was probably both: certainly capable of hunting, but with a body that big, it's hard to imagine they would turn down a free meal!

Image (previous):
Tyrannosaurus rex
Late Cretaceous Period
(66 million years ago)
Western North America

Catching A Meal

Generally speaking, animals prefer to not be eaten and have the tools to help them avoid that fate. Turkeys have excellent eyesight, rats have impeccable hearing, and gazelles will use their lightning speed and agility to outrun their would-be captors. Bears can smell both prey and other predators from miles away. Many animals have several of these adaptations, so a successful hunter has to bring a few tools to the trade as well.

Predators often have eyes set on the front of their faces, giving them better depth perception than animals with eyes set to the side. They also often rely on speed. For many mammals, like cheetahs, that speed comes from long bones in the legs and feet that help to increase their stride, or the distance between their steps. It is thought that the long legs of dromeosaurs were a similar adaptation, built for speed! Of course, it also helps to have sharp teeth, which sharks and *Majungasaurus* have in abundance! Hawks and eagles, like all birds, don't have teeth, but who needs them when you have razor-sharp claws at the ends of strong feet—just like *Velociraptor*!

This is the story of life: for every adaptation a prey species uses to avoid being hunted, the hunters have a tool to capture a meal. Undoubtedly, the dinosaurs lived through the same type of arms race.

Image:
Majungasaurus
Late Cretaceous Period
(67 million years ago)
Madagascar

Jason C. Poole
majungasaurus

Generalists vs Specialists

Animals living today display a wide variety of different feeding strategies. Many species are generalists and will eat nearly anything around them. Many bears and primates have teeth and appetites that allow them to eat meat while subsisting on a diet of nuts and berries for long portions of the year.

Other species have much more specialized diets and those feeding strategies are far more complicated than simply "carnivore or herbivore." Bison aren't "just" herbivores—they graze specifically on grasses. Deer don't just eat plants—they browse on the leaves, shoots, and fruit of shrubs. Vultures only scavenge already-dead animals, and predators hunt for fresh meat. But it gets even more specialized than that. Some predators, called piscivores, only eat fish. The giant panda takes specialized feeding to the extreme; they almost *exclusively* eat bamboo.

There is ample evidence that many species of dinosaurs and other ancient, extinct animals had specific feeding behaviors as well. The long, narrow snout of *Spinosaurus* suggests that it was a **piscivore**, and the short, blunt teeth of a mosasaur called *Globidens* were probably only effective for crushing the shells of clams, oysters, and ammonites. Both the medium-sized pterosaur, *Dimorphodon,* and the chicken-sized dinosaur, *Albertonykus*, had sharp, pointed front teeth like many bats, suggesting that they were **insectivores**, or insect eaters.

Image:
Pachycephalosaurus
Late Cretaceous Period
(70-66 million years ago)
Western North America

2021
Pachycephalosaurus
"Stygimoloch"

Image:
Albertonykus
Late Cretaceous Period
(68 million years ago)
Western North America

Piscivory

Spinosaurus was first found in Egypt in 1912, but paleontologists didn't know a lot about the species because the skeleton, which was only partially complete to begin with, was accidentally destroyed during World War II. Based on those original bones and a few others, though, we knew that it had a long, narrow snout, and a tall, thin sail along the length of its back. Paleontologists just assumed that the rest of the body looked similar to other large familiar theropods, like *Allosaurus* and *Tyrannosaurus rex*.

Almost exactly 100 years later paleontologists finally found a new, more complete skeleton in Morocco. It showed us that *Spinosaurus* actually wasn't much like those predators, or any other dinosaur, really. *Spinosaurus* had short hind legs that helped it live a mostly aquatic lifestyle—something we've never seen before among the Ruling Reptiles! And that long, narrow snout—does that remind you of any living animals today? Crocodiles and alligators of course! That's probably not just a coincidence; those elongate, tooth-filled faces are perfectly adapted to catching their favorite fishy prey!

The story of *Spinosaurus* is a great reminder of two important scientific principles: 1) that science is a process of changing our understandings based on new evidence, and; 2) that biology and behavior are always linked!

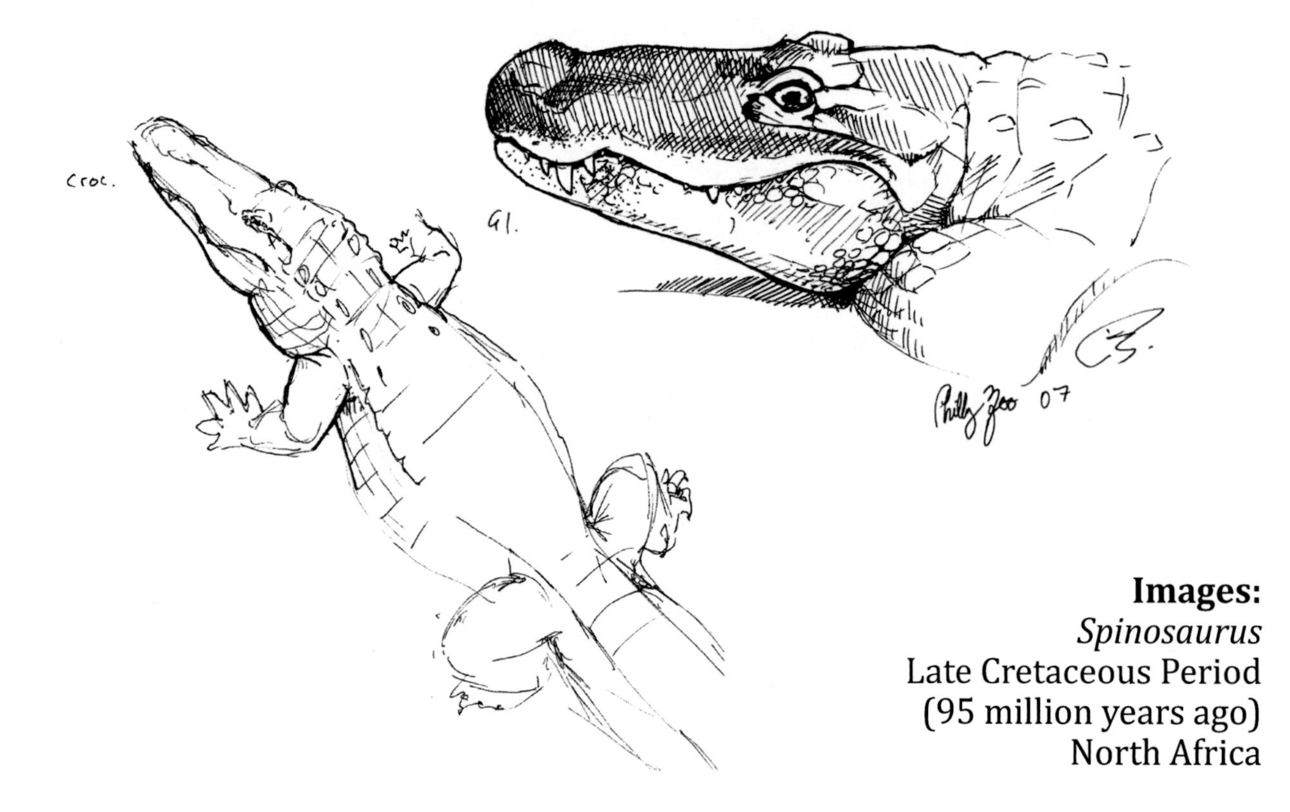

Images:
Spinosaurus
Late Cretaceous Period
(95 million years ago)
North Africa

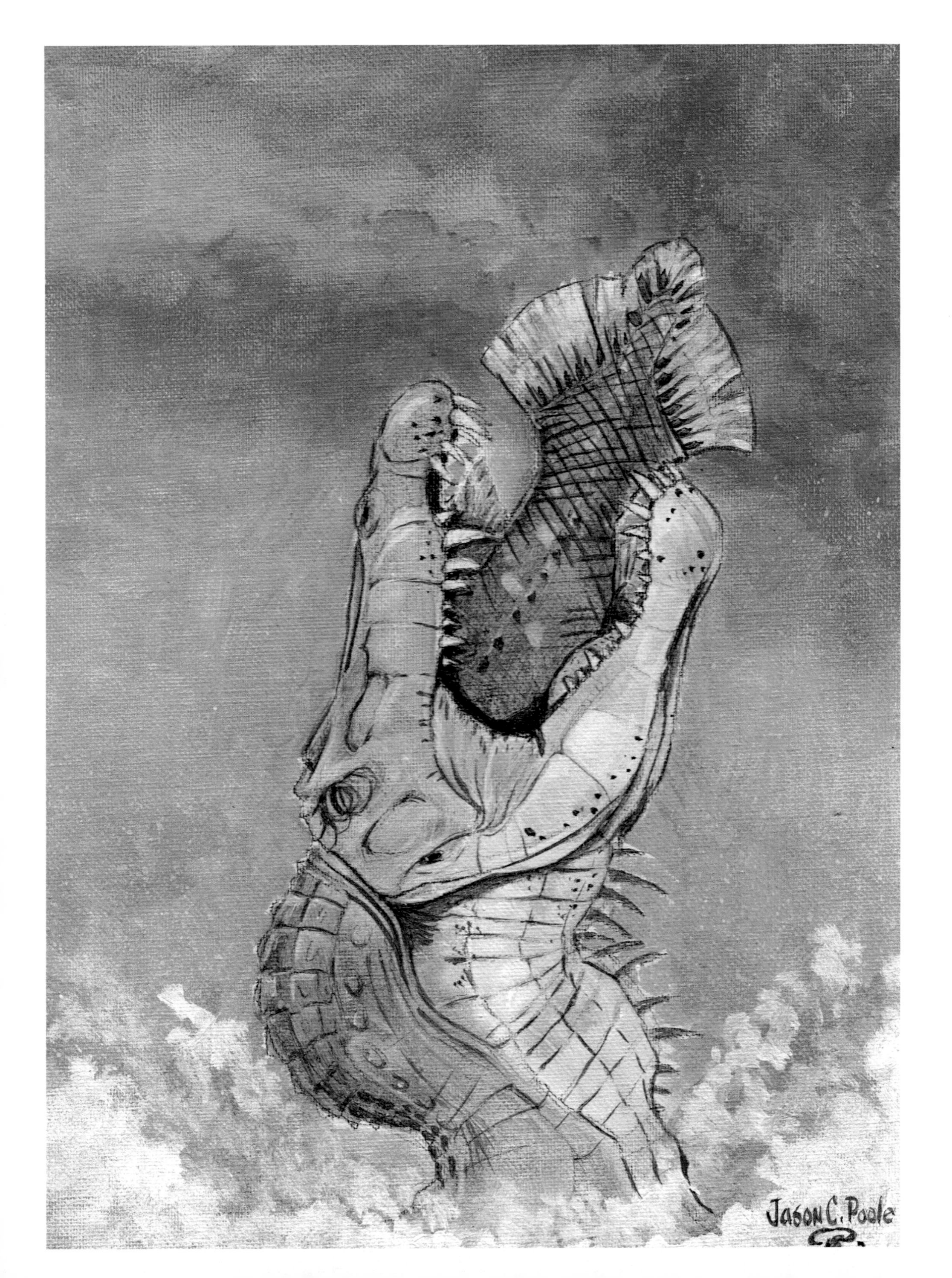
Jason C. Poole

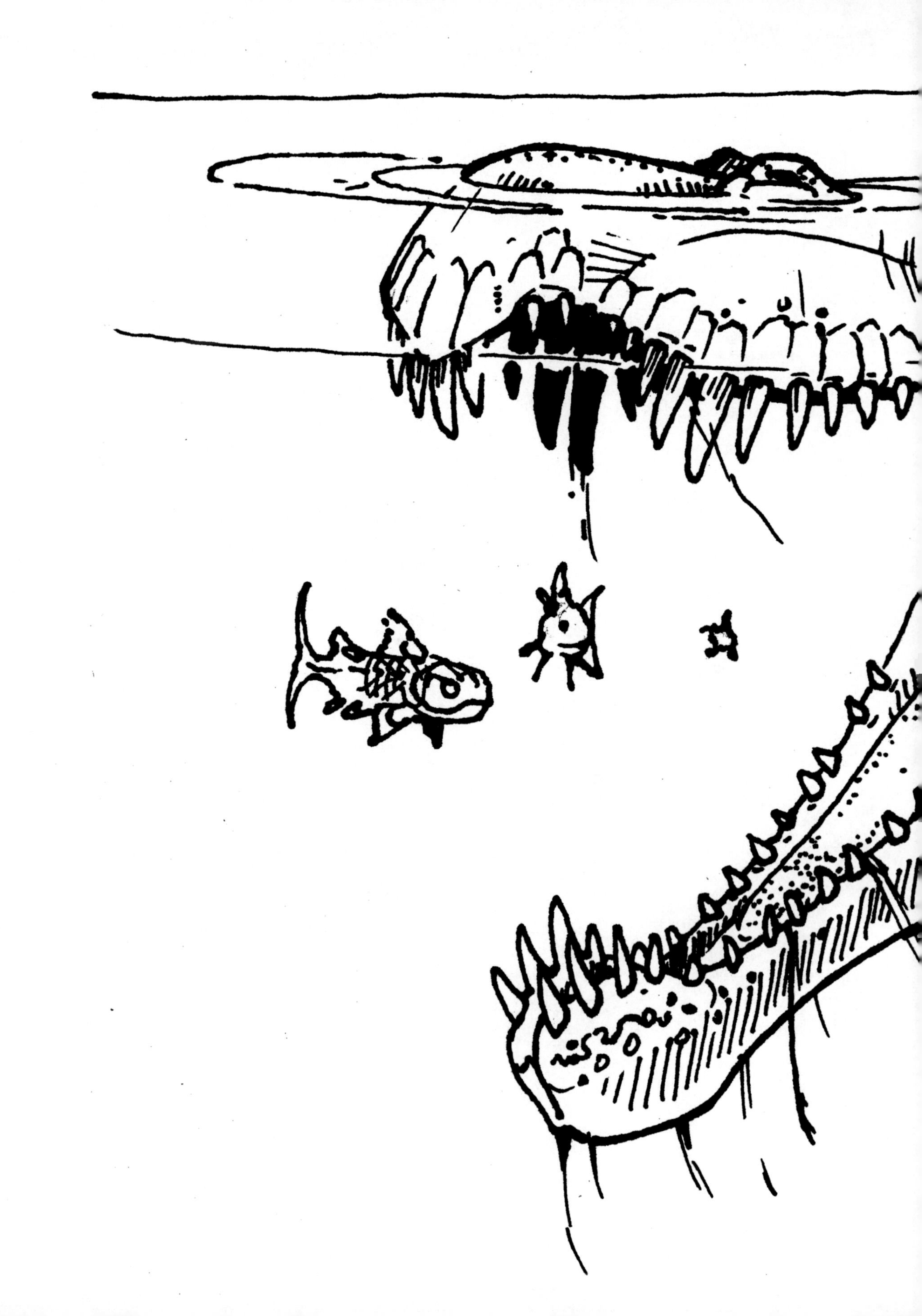

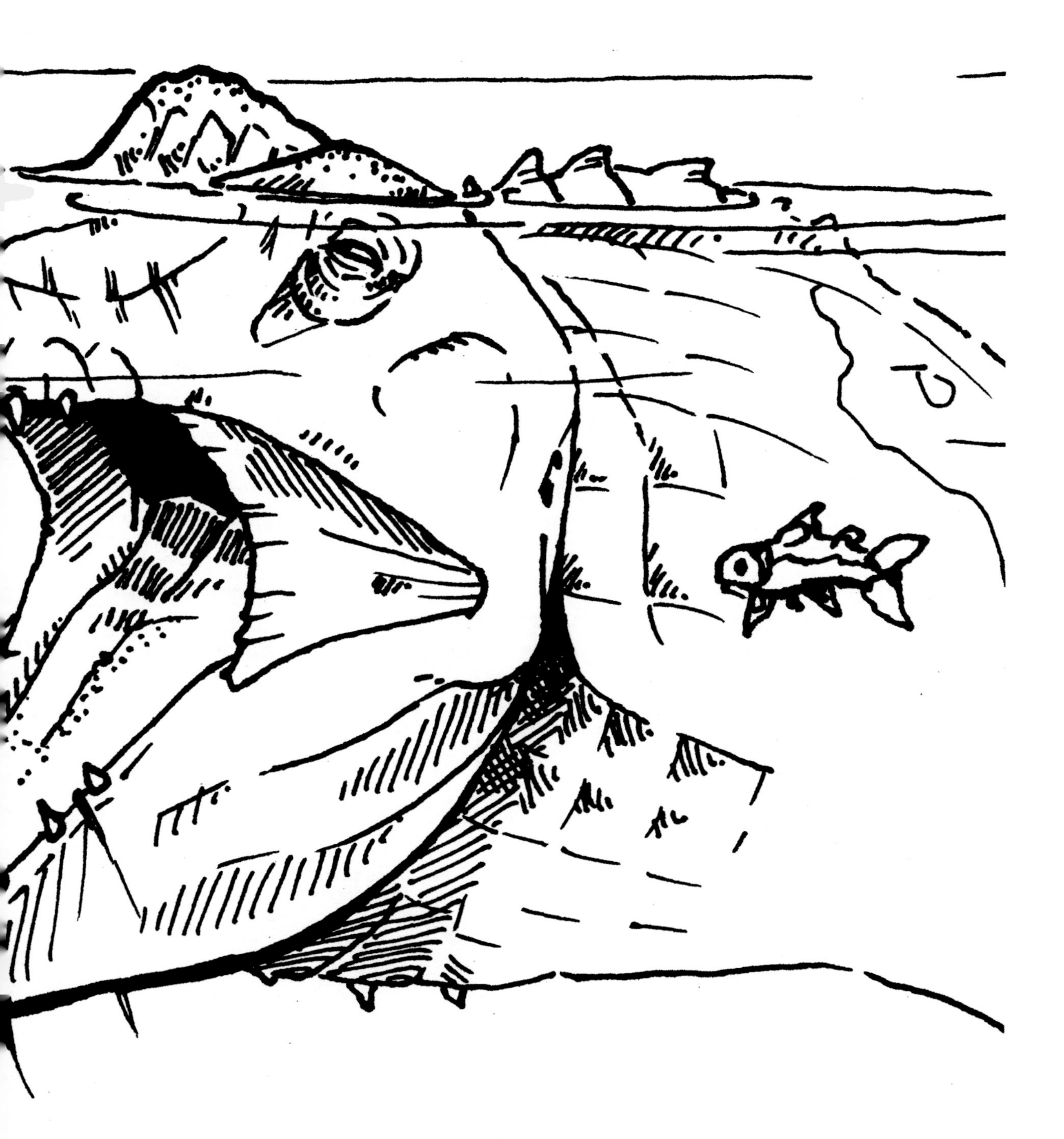

Pack Hunting

Most predators are loners, at least when it comes to hunting. Surprise is a primary tool in their arsenal, and it's difficult to sneak up on an unsuspecting victim if there are several helpers. Cheetahs and jaguars, owls, and snakes all hunt alone, relying on surprise and bursts of speed to capture their prey. But that's not the only way to get the job done.

Plenty of other predators don't care about surprising their future victims at all. Instead, their primary tactic is just the opposite—let the prey know they're coming and overwhelm them with greater numbers, *and* strategy. Wolves, African wild dogs, dolphins, chimpanzees, and even army ants hunt in cooperative groups, which is a fascinating and complicated behavior that requires foresight, planning, and advanced communication skills, all of which are rare in the animal kingdom.

It's quite likely that some theropods also cooperatively hunted in packs. Many species, especially raptors like *Coelophysis* and *Anzu*, had very delicate skeletons of thin, hollow bones that could be broken easily in a fight with larger prey. It is difficult to understand how they may have attacked prey successfully and safely without cooperatively hunting in packs.

Image:
Coelophysis
Late Triassic Period
(200 million years ago)
Western North America

2021
Jason C. Poole
Coelophysis

Raptorex

Play

After all the time spent finding and eating food, avoiding being eaten, and finding mates, you might think there isn't much time left for anything else except sleep. And yet, many species find time to have some fun. It turns out, playing is actually an important activity.

Primates, like ourselves, are the most playful creatures in the animal kingdom. Playing with siblings and other members of your species helps to strengthen relationships and family bonds. Some species of deer, zebras, and other hoofed animals run and leap into the air, especially when they are young; a behavior that helps to build strength and agility. The young of many predators, too, spend a lot of time wrestling each other, or "attacking" sticks, rocks, or anything else they can find. Young cats—from house kittens to lions—are famous for this kind of rambunctious play, as are puppies, bear cubs, and even river otters. This form of playing helps to build strength, coordination, and experience; all very helpful skills when they are older, and the attacks are a matter of life or death!

Because play is purely a behavior, it is difficult to imagine any evidence of dinosaur play being preserved in the fossil record. But because it is such an important activity seen across so much of the animal kingdom today, it is safe to assume dinosaurs enjoyed some leisure time once in a while, and in many of the same ways—play fighting among rambunctious young predators and running and wrestling among juvenile herbivores.

Image (previous):
Raptorex
Late Cretaceous Period
(70 million years ago)
Asia

Summary

As a scientist and nature lover, I want to know everything I possibly can about... well, just about everything in nature. It can be very frustrating to think about how much there is about the lives of dinosaurs that we may never know.

On the other hand, that's one of the wonderful things about dinosaur paleontology, and all of science. Our curiosity—that drive to know—pushes us to find creative new ideas that lead to innovative and ingenious solutions to seemingly insurmountable problems. There are many things that we thought just 15 or 20 years ago would be forever unknowable but are now some of our most exciting discoveries: skin pigment patterns, preserved soft tissues, vocalizations, and even migration rooutes, are now at the forefront of our studies. What's next? Colors? DNA? There has never been a more exciting time to study dinosaurs. We've never before had so many brilliant scientists working on so many fascinating questions. Will we ever know everything there is to know about the dinosaurs' world and lives? Definitely not, but we're learning more every day, and in turn, learning more about modern nature in the process. And just as importantly, we're having a lot of fun doing it!

Image:
Dilophosaurus
Early Jurassic Period
(193 million years ago)
Western North America

Jason C. Poole
©2021
Dilophosaurus

Image:
Morrison Ecosystem
Late Jurassic Period
(150 million years ago)
Western North America

Glossary:

Bonebed: Any layer of rock of geological deposit containing a great number of fossils bones.

Braincase: The part of the skull that forms a protective case around the brain.

Browser: An herbivorous animal that primarily eats leaves, soft shoots, and fruits from woody plants.

Carnivore: An animal that solely eats meat.

Carrion: Decaying flesh of dead animals.

Echolocation: The use of sounds, especially echoes, to locate food and navigate, often at night.

Ecosystem: All of the organisms and the physical environment in which they interact.

Ethology: The study of modern animal behavior.

Extinction: The complete and permanent termination of a species or group of species of organisms following the death of the last individual.

Flagellicaudata: A group of sauropod dinosaurs characterized by long, thin, whip-like tails, like *Diplodocus*.

Flocking: The behavior of a large group of animals, especially birds, sheep, and goats, while foraging or moving together.

Gastrolith: Stones picked up and swallowed by animals to help digest their food.

Grazer: An animal that primarily eats grass.

Herbivore: An animal that exclusively eats plants.

Infrared: Light that has wavelengths longer than visible light. Animals that use infrared vision see heat rather than colors as we humans do.

Insectivore: Animals that primarily eat insects.

Insulator: A material or substance that reduces heat loss.

Intra-species: Between members of the same species.

Keratin: A biological material that makes up scales, hair, feathers, hooves, claws, and fingernails.

Megafauna: Large animals, typically at over 100 lbs.

Niche: The role an organism has within its ecosystem.

Obligate Scavenger: Animals that exclusively eat dead and decaying organisms.

Osteoderms: Bones that grow directly out of the skin.

Pair Bonding: The behavior of some animals to mate with a single other individual for extended periods of time, and sometimes for life.

Piscivore: An animal that primarily eats fish.

Preening: A behavior where birds clean and maintain their feathers with their beaks.

Quill knob: The attachment sites for flight feathers on the bone.

Ruminant: A group of herbivorous, hoofed mammals that include deer, antelope, and goats.

Rut: The mating season among some mammals.

Subterranean: Underground.

Symbiosis: A close and long-term biological interaction between two different species.

About the Author:

Jason P. Schein is a writer, naturalist, explorer, conservationist, science-evangelist, and a paleontologist. He has participated in paleontological expeditions in Argentinean Patagonia and across the U.S., and continues to lead expeditions in Montana and Wyoming as the founder of the Bighorn Basin Paleontological Institute. He holds a B.S. and an M.S. in geology from Auburn University and has taught geology and paleontology at universities in Alabama and Philadelphia, Pennsylvania. During the offseason he lives in Philadelphia, where he fills his time by sharing his love of the natural world with his family.

About the Illustrator:

Jason Poole hatched in 1970 someplace in the eastern United States where he was raised by squirrels. His interest in art was spawned as he learned to read from discarded comic books in the mean streets of Philadelphia. To this day the dynamic line art of comic books has left an impression on "Mr. Poole" as he is called by his jungle friends. Mr. Poole has produced art for National Geographic and National Geographic World magazines, several scientific publications, and museums. Jason enjoys long walks in the badlands of Montana and Wyoming and spending time with his offspring and life mate.

Bighorn Basin Paleontological Institute:

The Bighorn Basin Paleontological Institute is a nonprofit 501(c)(3) organization dedicated to paleontology and earth science research, education, and outreach. We harness the universal appeal of dinosaurs and paleontology to engage people of all interest levels and backgrounds to promote the earth and natural sciences. Through exploration and discovery, our scientists, citizen scientists, and volunteers contribute to our understanding of the prehistoric worlds and the planet we inhabit.